Student Workbook

by
Robert H. Marshall
Allen B. Rosskopf

AGS Publishing
Circle Pines, MN 55014-1796
800-328-2560

© 2004 AGS Publishing
4201 Woodland Road
Circle Pines, MN 55014-1796
800-328-2560 • www.agsnet.com

AGS Publishing is a trademark and trade name of American Guidance Service, Inc.

All rights reserved, including translation. No part of this publication may be reproduced or transmitted in any form or by any means without written permission from the publisher.

Printed in the United States of America

ISBN 0-7854-3637-5

Product Number 93943

A 0 9 8 7 6 5 4 3 2 1

Table of Contents

Workbook Activity	1	Earth Science
Workbook Activity	2	Understanding Maps
Workbook Activity	3	Topographic Maps: Terms Review
Workbook Activity	4	The Earth's Features
Workbook Activity	5	The Earth's Rotation and Time
Workbook Activity	6	A Grid System on a Map
Workbook Activity	7	Latitude
Workbook Activity	8	Longitude
Workbook Activity	9	A Grid System on a Globe: Terms Review
Workbook Activity	10	The Effect of Gravity
Workbook Activity	11	The Earth's Movement in Space
Workbook Activity	12	The Moon's Movement in Space
Workbook Activity	13	The Moon's Surface: Terms Review
Workbook Activity	14	The Solar System
Workbook Activity	15	The Inner Planets
Workbook Activity	16	The Outer Planets
Workbook Activity	17	Other Objects in the Solar System: Terms Review
Workbook Activity	18	Stars
Workbook Activity	19	Distances to Stars
Workbook Activity	20	The Life of a Star
Workbook Activity	21	Groups of Stars: Terms Review
Workbook Activity	22	Matter
Workbook Activity	23	The Smallest Parts of Matter
Workbook Activity	24	Compounds and Mixtures: Terms Review
Workbook Activity	25	Minerals
Workbook Activity	26	Properties Used to Identify Minerals
Workbook Activity	27	Other Physical Properties of Minerals
Workbook Activity	28	Common Uses of Minerals: Terms Review
Workbook Activity	29	Rocks and Rock Types

Workbook Activity	30	Igneous Rocks
Workbook Activity	31	Sedimentary Rocks
Workbook Activity	32	Metamorphic Rocks
Workbook Activity	33	The Rock Cycle: Terms Review
Workbook Activity	34	Gases in the Atmosphere
Workbook Activity	35	Layers of the Atmosphere
Workbook Activity	36	Clouds
Workbook Activity	37	Precipitation
Workbook Activity	38	Wind Patterns: Terms Review
Workbook Activity	39	Weather Conditions and Measurements
Workbook Activity	40	Weather Patterns and Predictions
Workbook Activity	41	Storms
Workbook Activity	42	World Climates: Terms Review
Workbook Activity	43	The Water Cycle
Workbook Activity	44	Sources of Fresh Water
Workbook Activity	45	Oceans: Terms Review
Workbook Activity	46	Weathering
Workbook Activity	47	Erosion Caused by Water
Workbook Activity	48	Erosion Caused by Glaciers
Workbook Activity	49	Erosion Caused by Wind and Gravity: Terms Review
Workbook Activity	50	Movement of the Earth's Crust
Workbook Activity	51	Volcanoes
Workbook Activity	52	Mountains
Workbook Activity	53	Earthquakes: Terms Review
Workbook Activity	54	The Rock Record
Workbook Activity	55	The Ages of Rocks and Fossils
Workbook Activity	56	Eras in the Geologic Time Scale: Terms Review

Name _____ Date _____ Period _____ | Workbook Activity Chapter 1, Lesson 1 — 1

Earth Science

Directions Write the term from the Word Bank that best completes each sentence.

Word Bank				
astronomy	earth science	exploration robot	lens	oceanography
balloons	El Niño	geologist	meteorology	submersible
computer	environment	geology	metric system	weather

1. The study of the surface and inside of the earth is called _____.

2. The study of the earth's water is called _____.

3. The study of the earth's air and weather is called _____.

4. The study of objects in space is called _____.

5. Land, water, and air are everywhere on earth.
   These interact and affect the _____ where you live.

6. Scientists usually base their measurements on the _____.

7. A very simple tool that helps a geologist study rocks is a hand-held _____.

8. A(n) _____ is a complicated tool astronomers use to study rocks on Mars.

9. To study the bottom of an ocean, oceanographers may use a(n)
   _____.

10. A(n) _____ is a tool that helps scientists store and analyze information as well as communicate with other scientists.

11. Communities depend on a knowledge of _____ to determine the best ways to use the land and its resources.

12. A(n) _____ is a scientist who might locate the oil that is used to power buses and cars.

13. Farmers rely on meteorologists to give them accurate reports on the _____.

14. Scientists who study the air often depend on _____ to measure weather conditions high above the earth.

15. A strong, warm climate system that repeats itself every three to seven years is called
    _____.

Earth Science

Name _____ Date _____ Period _____ | **Workbook Activity**
Chapter 1, Lesson 2 — **2**

Understanding Maps

Directions Match each term with its definition. Write the letter of the correct definition on the line.

_____ 1. compass directions
_____ 2. compass rose
_____ 3. legend
_____ 4. map
_____ 5. model
_____ 6. scale
_____ 7. symbol

A drawing that shows the earth's surface from above
B list of map symbols and their meanings
C map symbol showing major compass directions
D object that shows how something looks or works
E relationship between actual distance and distance shown on a map
F north, south, east, and west
G word, shape, or color that represents something else

Directions Compare or contrast each pair of terms.

8. map—model
   How they are alike _____
   _____
   _____

9. compass rose—scale
   How they are different _____
   _____
   _____

10. symbol—legend
    How they are different _____
    _____
    _____

Earth Science

Name _____ Date _____ Period _____

Workbook Activity
Chapter 1, Lesson 3
3

Topographic Maps: Terms Review

Directions Match each field of science with its description. Write the letter on the line.

_____ 1. Earth Science			
_____ 2. Geology	_____ 3. Oceanography	_____ 4. Meteorology	_____ 5. Astronomy

A the study of the earth's air and weather

B the study of the earth's oceans

C the study of outer space and objects in it

D the study of the earth's land, water, air, and outer space

E the study of the solid parts of the earth

Directions Write the term from the Word Bank that best completes each sentence.

Word Bank
| compass rose | contour interval | legend | metric system | submersible |
| contour lines | hachure | map | scale | unit |

6. A _____ is the vertical distance between contour lines on a topographic map.

7. On a topographic map, a _____ is used to show a depression.

8. A _____ shows direction on a map.

9. A ratio can be used for a map _____.

10. A _____ is a drawing of part of the earth's surface as seen from above.

11. A _____ lists map symbols and their meanings.

12. Scientists most often use the _____ of measurement.

13. On a topographic map, _____ connect points of equal elevation.

14. A meter is a metric _____ of length.

15. Oceanographers might use a _____ to explore underwater.

Earth Science

Name _____ Date _____ Period _____ | Workbook Activity
Chapter 2, Lesson 1
4

The Earth's Features

Directions 1–11. Look at the map of the earth. Label each of the seven continents on the map. Then label each of the four oceans.

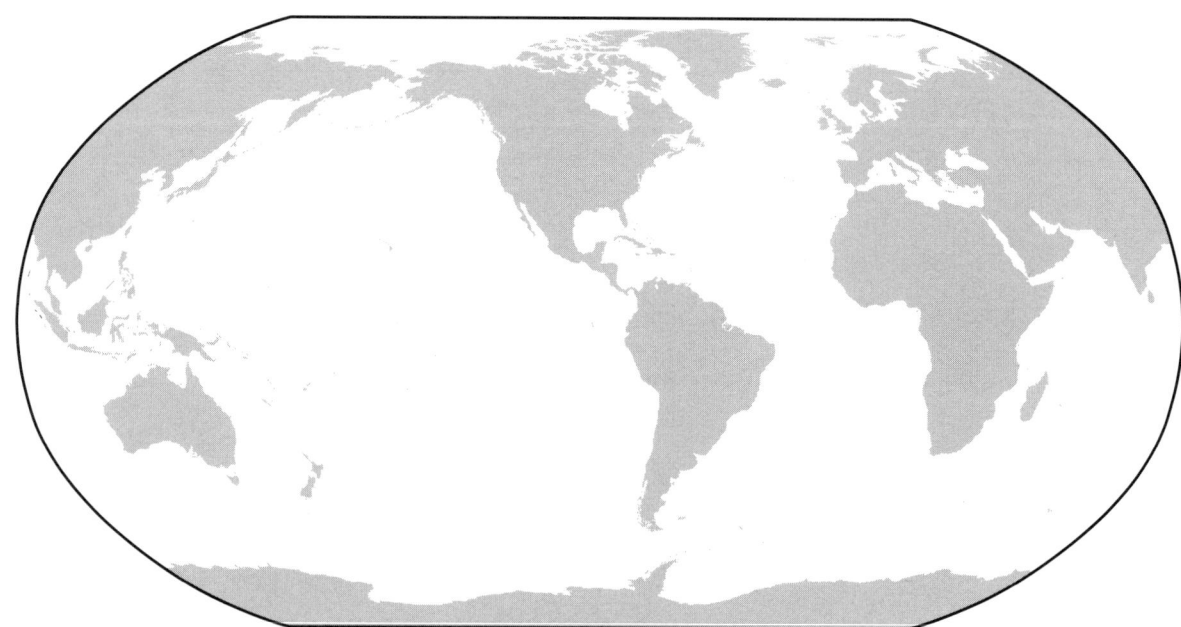

Directions Answer the questions.

12. Which continent is the smallest? _____

13. Which ocean is the largest? _____

14. In kilometers, how much larger is the distance around the earth from east to west than the distance from north to south? _____

15. What percent of the earth's land area does North America cover? _____

16. Which ocean is the deepest? _____

17. Why might the earth's oceans be described as one huge ocean?
_____

18. How is the earth similar to and different from a spinning top?
_____

19. Besides distance, why can't you see across an ocean to land on the other side?
_____

20. Use the circle to draw a circle graph that shows how much of the earth is water and how much is land.

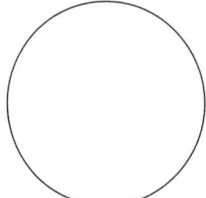 Earth Science

Name	Date	Period	**Workbook Activity**
			Chapter 2, Lesson 2 5

The Earth's Rotation and Time

Directions For each pair of cities below, use the time zone map to figure out the time in the second city. Write the time on the line. Be sure to include A.M. or P.M.

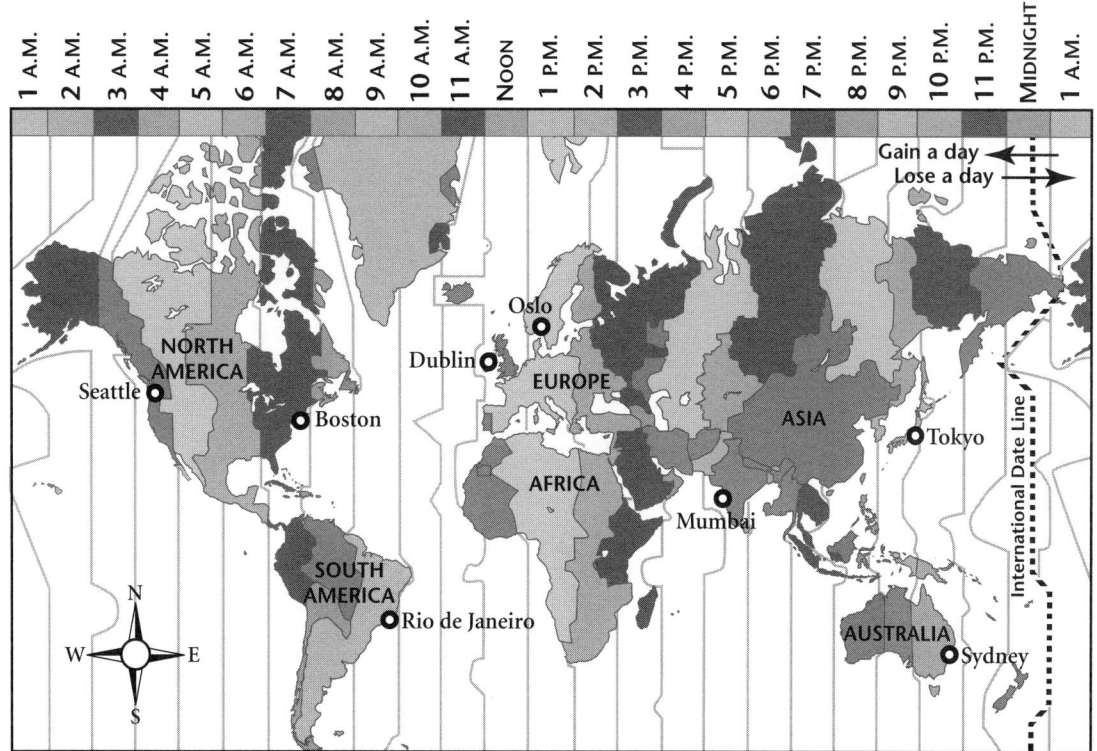

1. 4:00 P.M.   Boston, Massachusetts
   _____   Sydney, Australia

2. 1:00 A.M.   Rio de Janeiro, Brazil
   _____   Tokyo, Japan

3. 11:00 P.M.   Mumbai, India
   _____   Dublin, Ireland

4. 8:00 A.M.   Oslo, Norway
   _____   Seattle, Washington

5. 12:00 NOON   Seattle, Washington
   _____   Rio de Janeiro, Brazil

6. 2:00 P.M.   Sydney, Australia
   _____   Mumbai, India

7. 9:00 P.M.   Oslo, Norway
   _____   Dublin, Ireland

8. 2:00 P.M.   Mumbai, India
   _____   Rio de Janeiro, Brazil

Directions Answer the questions.

9. In hours, what is the time difference between Boston and Rio de Janeiro? _____

10. If you traveled from Oslo to Tokyo, would you move your watch forward or backward?

_____

Earth Science

Workbook Activity
Chapter 2, Lesson 3
6

A Grid System on a Map

Directions Use the map to answer the questions below.

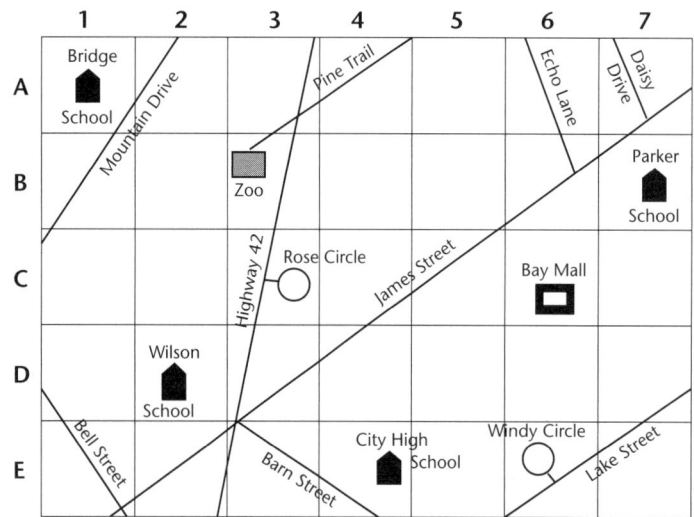

1. In which block is most of Echo Lane? _____
2. In which block is the school that is closest to Mountain Drive? _____
3. In which block does Pine Trail cross Highway 42? _____
4. In which block is City High School? _____
5. In which block does Echo Lane meet James Street? _____
6. In which block is Rose Circle? _____
7. In which block is Bay Mall? _____
8. Which blocks contain parts of Bell Street? _____
9. What school is located in block B7? _____
10. In which block is the zoo? _____
11. To get to the corner of Barn Street and James Street from the corner of Echo Lane and James Street, what blocks would you travel through? _____
12. In which block does James Street cross Bell Street? _____
13. Michael wants to open a restaurant on a main street close to Bay Mall. Identify two areas by block names that would fit this description. _____
14. Jackie lives on Pine Trail near the zoo. What blocks would she travel through to take Highway 42 and Barn Street to City High School? _____
15. The Perezes are considering buying one of two houses. One is on Barn Street. The other is at the corner of Lake Street and Windy Circle. Which is closer to Wilson School? _____

Earth Science

Name _____ Date _____ Period _____ | **Workbook Activity** Chapter 2, Lesson 4 — 7

Latitude

Directions Match the lettered features in the diagram with the descriptions below. Write the correct letter on each line.

1. 30°N _____
2. 30°S _____
3. 60°N _____
4. 60°S _____
5. equator _____
6. North Pole _____
7. South Pole _____

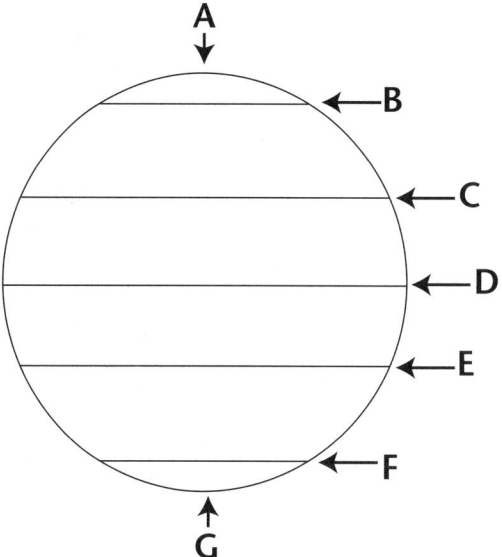

Directions Use the map of California to answer the questions.

8. What is the approximate latitude of San Francisco? _____
9. What is the approximate latitude of Fresno? _____
10. What is the approximate latitude of Los Angeles? _____
11. What is the approximate latitude of San Diego? _____
12. What is the approximate latitude of the northern border of California? _____

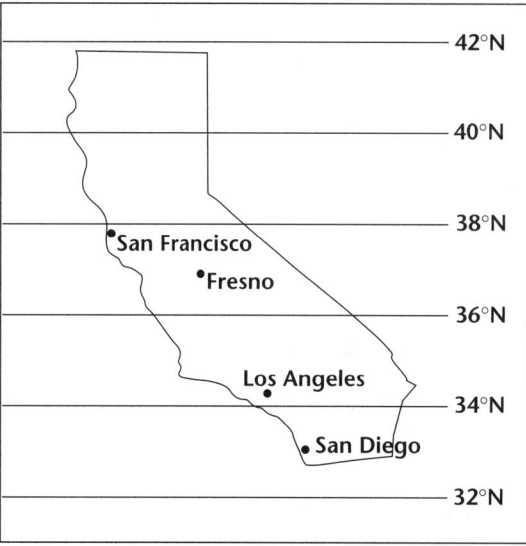

Directions Answer the questions.

13. What is another name for lines of latitude? How do you think they got this name?
_____

14. What is the latitude of the equator? _____

15. What is the northernmost latitude on the earth? _____

Earth Science

Name _____ Date _____ Period _____

Workbook Activity
Chapter 2, Lesson 5
8

Longitude

Directions Use the map to estimate the longitude of each city below. Write the longitude for each city on the line.

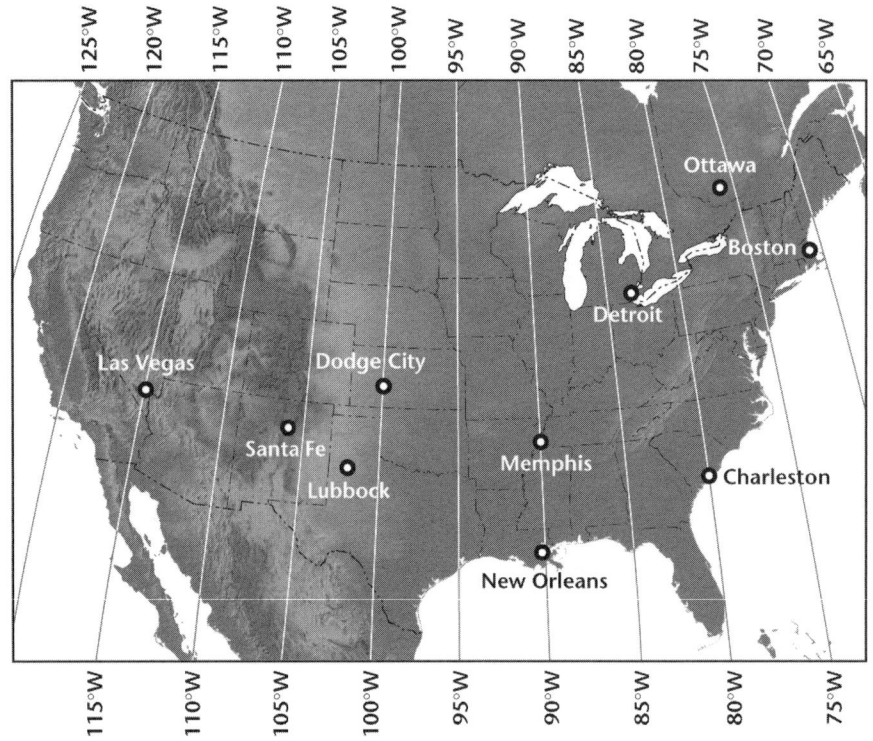

1. Charleston, South Carolina _____
2. Santa Fe, New Mexico _____
3. Memphis, Tennessee _____
4. Boston, Massachusetts _____
5. Dodge City, Kansas _____
6. Lubbock, Texas _____
7. New Orleans, Louisiana _____
8. Detroit, Michigan _____
9. Las Vegas, Nevada _____
10. Ottawa, Canada _____

Directions The longitude is given for each city listed below. Arrange the cities in order from west to east.

| Green Bay, Wisconsin, 88°W |
| Columbus, Ohio, 83°W |
| Pierre, South Dakota, 100°W |
| Amarillo, Texas, 102°W |
| St. Louis, Missouri, 90°W |

11. _____
12. _____
13. _____
14. _____
15. _____

Earth Science

A Grid System on a Globe: Terms Review

Directions Read the clues to complete the puzzle.

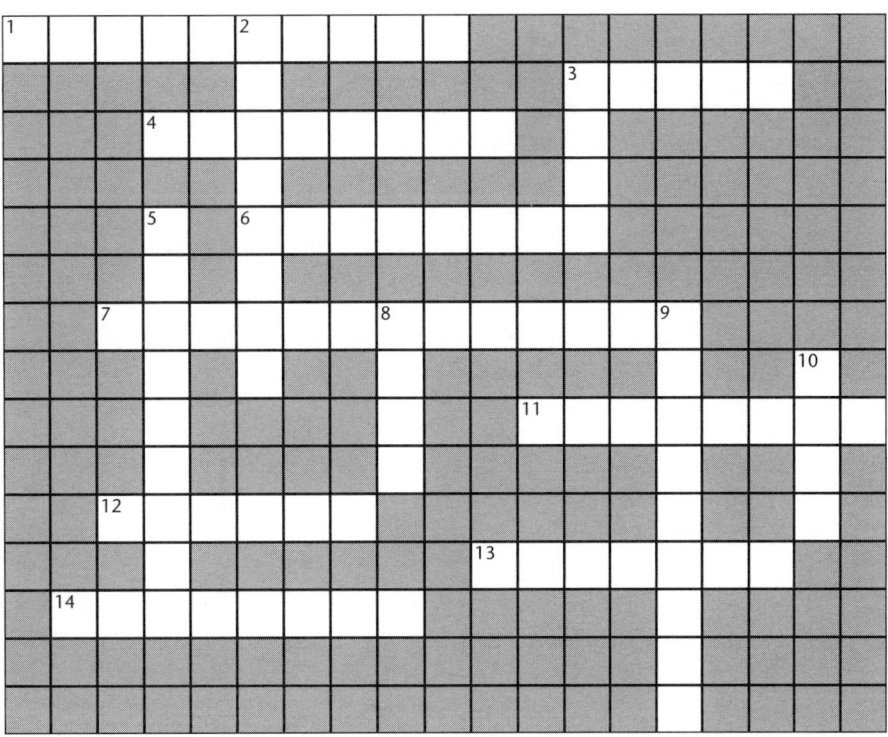

Across

1. half of the earth
3. _____ meridian: line of 0° longitude
4. line of longitude
6. angle that describes the distance north or south of the equator
7. _____ date line: marks the place on the earth where each new day begins
11. _____ time zone: an area that has the same clock time
12. unit for measuring angles in a circle or sphere
13. line of 0° latitude
14. spinning of the earth

Down

2. line of latitude
3. point that the earth's axis passes through
5. one of the seven major land areas of the earth
8. imaginary line through the earth
9. angle that describes the distance east or west of the prime meridian
10. set of horizontal and vertical lines on a map

Name _____ Date _____ Period _____ | **Workbook Activity** Chapter 3, Lesson 1 | **10**

The Effect of Gravity

Directions Study the diagram. Write a sentence below describing Orbit A. Write another sentence describing Orbit B.

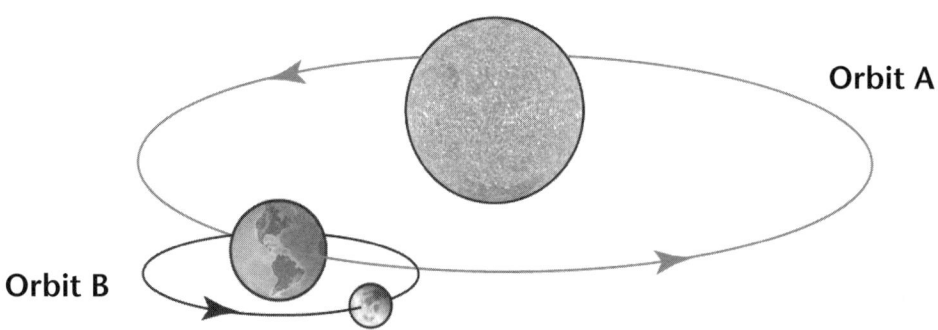

1. Orbit A: _____
   _____

2. Orbit B: _____
   _____

Directions Gravity holds planets and satellites in orbit. Write *larger* or *smaller* to complete each sentence.

3. The sun has a _____ gravitational pull than the earth.

4. The earth has a _____ gravitational pull than the moon.

5. The moon has a _____ gravitational pull than the sun.

Directions Write the term from the Word Bank that best completes each sentence.

6. Gravity is a force that causes two objects to _____ on each other.

7. The force of gravity between two objects depends on the _____ of each object, or how much matter each contains.

8. How much two objects pull on each other also depends on their _____ from each other.

9. The moon moves in a curved path, or _____, around the earth.

10. The earth orbits the _____.

Word Bank
distance
mass
orbit
pull
sun

Earth Science

Name	Date	Period	**Workbook Activity**
			Chapter 3, Lesson 2 — 11

The Earth's Movement in Space

Directions Write the term from the Word Bank that best completes each sentence.

1. One orbit around an object is one _____.
2. The earth revolves around the sun once a _____.
3. A sphere spins one time; this equals one _____.
4. The earth rotates on its axis once a _____.
5. Seasons occur because the earth revolves and is _____ on its axis.
6. The angle of tilt of the earth's _____ is exactly 23.5 degrees.
7. The part of the earth that is tilted toward the sun is having _____.
8. This happens because the sun's rays strike the earth more _____ then.
9. The part of the earth that is tilted away from the sun has _____.
10. There is less heat when the sun's rays strike the earth at an _____.

Word Bank
angle
axis
day
directly
revolution
rotation
summer
tilted
winter
year

Directions Use the diagram on page 61 of the textbook to complete the table below. For angle of light, write *direct*, *in between*, or *indirect*. Parts of the chart have been filled in.

Hemisphere	Date	Angle of Light	Season
Northern	December 21	**11.**	**12.**
Southern	December 21	**13.**	**14.**
Northern	March 21	**15.**	**16.**
Southern	March 21	in between	fall
Northern	June 21	direct	summer
Southern	June 21	indirect	winter
Northern	September 23	**17.**	**18.**
Southern	September 23	**19.**	**20.**

Earth Science

Name _____ Date _____ Period _____ | **Workbook Activity**
Chapter 3, Lesson 3 — **12**

The Moon's Movement in Space

Directions When you compare and contrast things, you tell how they are alike and how they are different. Compare and contrast each pair of terms.

revolution—rotation

1. Alike _____

2. Different _____

earth—moon

3. Alike _____

4. Different _____

lunar eclipse—solar eclipse

5. Alike _____

6. Different _____

new moon—full moon

7. Alike _____

8. Different _____

high tide—low tide

9. Alike _____

10. Different _____

Directions Match the term with its definition. Write the letter of the correct definition on the line.

____ 11. full moon

____ 12. lunar eclipse

____ 13. phases of the moon

____ 14. tides

____ 15. new moon

A when the moon passes through the earth's shadow

B when the earth is between the sun and the moon; the moon is fully lit

C when the moon is between the sun and the earth; the moon looks dark

D changes in the moon's appearance as it orbits

E regular rising and falling of the earth's oceans

Earth Science

The Moon's Surface: Terms Review

Directions Read the clues to complete the puzzle.

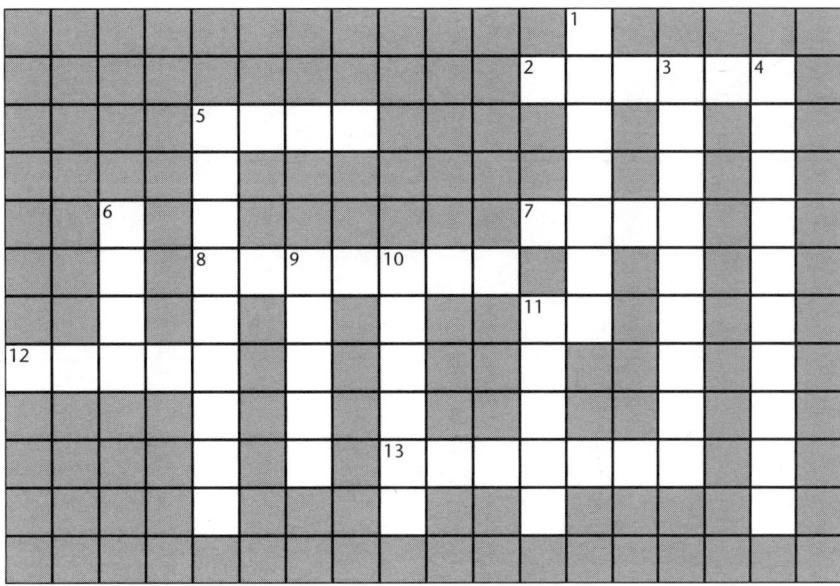

Across

2. feature caused by an object hitting the surface of the moon
5. amount of matter that an object contains
7. rising or falling of the ocean level because of the moon's pull
8. oval shape of an orbit
11. belonging to me
12. ____ eclipse: when the moon casts a shadow on the earth
13. shadow cast by the earth or the moon

Down

1. force of attraction between any two objects
3. instrument that makes distant objects look brighter and bigger
4. movement of an object in orbit around another object in space
5. rock from space that hits the surface of a planet or moon
6. ____ moon: phase when the earth is between the sun and the moon
9. ____ eclipse: passing of the moon through the earth's shadow
10. changes in the way the moon appears as it orbits the earth
11. flat, low plains on the moon

Name _____ Date _____ Period _____

**Workbook Activity
Chapter 4, Lesson 1
14**

The Solar System

Directions When you compare and contrast things, you tell how they are alike and different. Compare and contrast each pair of terms.

star—planet

1. Alike _____

2. Different _____

planet—moon

3. Alike _____

4. Different _____

moon—star

5. Alike _____

6. Different _____

planet—sun

7. Alike _____

8. Different _____

Directions Write the term from the Word Bank that best completes each sentence.

Word Bank			
chromosphere	helium	nuclear reactions	three
cooler	mass	star	

9. The sun is a _____ because it makes its own energy and light.

10. The sun is made mostly of _____ and hydrogen.

11. The sun's high temperature is caused by _____ inside the sun.

12. The sun's atmosphere has _____ layers.

13. The middle layer of the sun's atmosphere is called the _____.

14. Because sunspots give off less energy, they are _____ than the rest of the sun.

15. The sun contains 99 percent of the _____ in the entire solar system.

Earth Science

Name _____ Date _____ Period _____ | **Workbook Activity** Chapter 4, Lesson 2 | **15**

The Inner Planets

Directions Look at the diagram. Then write the names of the four inner planets as headings in the chart below. Fill in the rest of the chart.

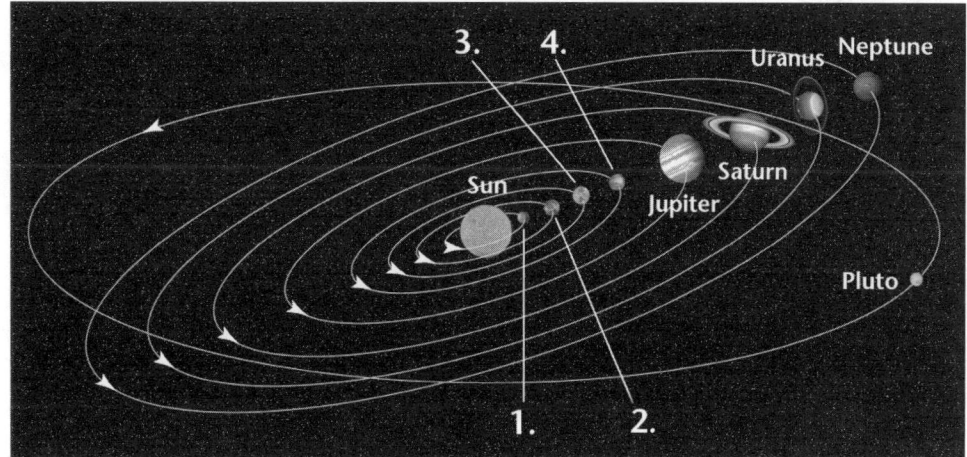

Planet Features	1.	2.	3.	4.
length of rotation	59 Earth days	5.	24 hours	6.
direction of rotation	west to east	7.	8.	west to east
temperature	9.	460°C	10.	colder than Earth
atmosphere	almost none	11.	12.	thinner atmosphere than Earth
surface	13.	plains, highlands, craters	liquid water, life	iron in rocks and soil
number of moons	14.	none	1	15.

Earth Science

Name _____ Date _____ Period _____ | Workbook Activity
Chapter 4, Lesson 3
16

The Outer Planets

Directions Look at the diagram. Then write the names of the five outer planets on the lines below.

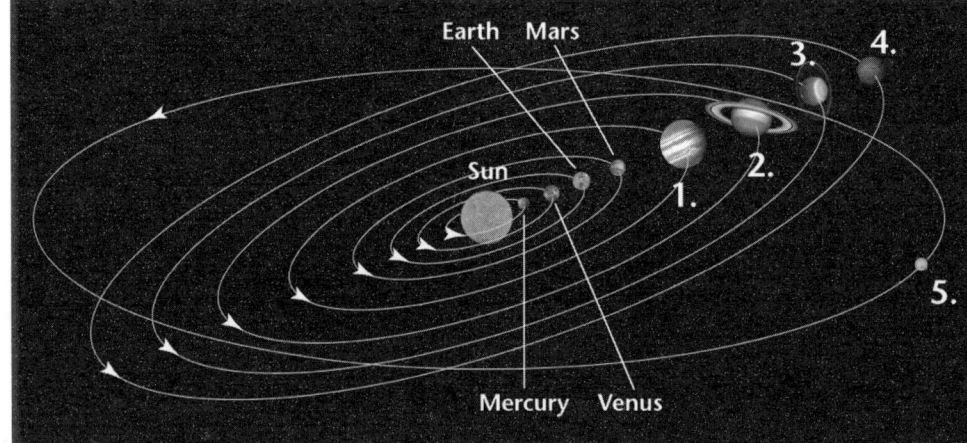

1. _____
2. _____
3. _____
4. _____
5. _____

Directions Write the answer to each question.

6. Which is the only outer planet that does not have a thick atmosphere? _____
7. Which planet is named after the Roman god of the sea? _____
8. Which planet is the largest? _____
9. Which planet rotates on its side? _____
10. Which is the second largest planet in the solar system? _____
11. Which planet has a moon called Charon? _____
12. Which planet takes 84 years to orbit the sun? _____
13. Which planet has a windstorm that has lasted at least 300 years? _____
14. Which planet has a feature called the Great Dark Spot? _____
15. Which planet has a moon called Titan? _____

Earth Science

Name _____ Date _____ Period _____ | Workbook Activity
Chapter 4, Lesson 4 — 17

Other Objects in the Solar System: Terms Review

Directions Match each clue with a term. Write the letter of the correct term on the line.

_____ 1. fastest-moving planet

_____ 2. object that makes its own light

_____ 3. planet named for the Roman god of war

_____ 4. envelope of gas surrounding an object in space

_____ 5. planet that rotates in the opposite direction from the others

_____ 6. cooler, darker area on the sun's surface

_____ 7. region between Mars and Jupiter where most asteroids are found

_____ 8. planet known for its rings

_____ 9. asteroid that enters Earth's atmosphere

_____ 10. ball of ice, rock, frozen gases, and dust

_____ 11. planet that rotates on its side

_____ 12. asteroid that hits the surface of a planet or moon

_____ 13. large object in space that orbits a star

_____ 14. object, smaller than a planet, that orbits a star

_____ 15. planet that is third from the sun

_____ 16. star and all of the objects that revolve around it

_____ 17. warming of the atmosphere because of trapped heat energy from the sun

_____ 18. largest planet

_____ 19. satellite that orbits a planet

_____ 20. greenish-blue planet with the Great Dark Spot

A asteroid
B asteroid belt
C atmosphere
D comet
E Earth
F greenhouse effect
G Jupiter
H Mars
I Mercury
J meteor
K meteorite
L moon
M Neptune
N planet
O Saturn
P solar system
Q star
R sunspot
S Uranus
T Venus

Earth Science

Name	Date	Period	Workbook Activity Chapter 5, Lesson 1 — 18

Stars

Directions The diagram shows the apparent magnitudes of six stars. Use the diagram to answer the questions below.

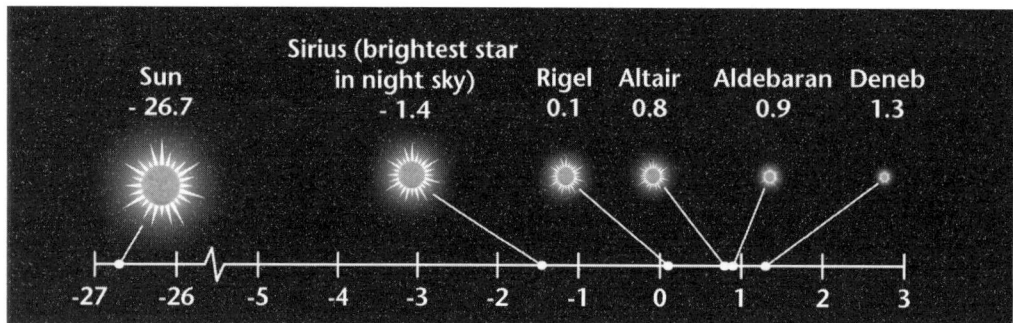

1. What is apparent magnitude? _____

2. Which star on the chart has the greatest apparent magnitude? _____

3. Is Rigel brighter or dimmer than Altair? _____

4. What is the apparent magnitude of the sun? _____

5. Which star is brighter: Sirius or Deneb? _____

Directions The table shows the average surface temperatures of different colors of stars. Use the table to answer the questions below.

6. Vega is a white star. What is its average surface temperature? _____

7. Arcturus is an orange-red star. Is it hotter or cooler than the sun? _____

Star Color and Temperature	
Color	Average Surface Temperature (°C)
blue-white	35,000
white	10,000
yellow	5,500
red	3,000

8. Is the surface temperature of a star higher or lower than the temperature inside the star? _____

Directions Answer the questions.

9. What is the name of the process that creates energy in stars? _____

10. How does this process make stars shine? _____

Earth Science

Name _____ Date _____ Period _____ | Workbook Activity
Chapter 5, Lesson 2
19

Distances to Stars

Directions Write the answer to each question.

1. What is the speed of light?
   _____

2. Does a light-year measure time, brightness, or distance?
   _____

3. What measurement does 1 light-year equal?
   _____

4. The star Procyon is 11 light-years away. What is its distance in kilometers?
   _____

5. The star Betelgeuse is 2,850 trillion kilometers away. What is its distance in light-years?
   _____

6. If you look at a star that is 14 light-years away, how long ago did the light you are seeing leave the star?
   _____

7. When you see light from the sun, how long ago did the light leave the sun?
   _____

8. How do astronomers determine how far away a star is by observing it once and then a few months later?
   _____

9. How do astronomers figure out the distance of stars that have shifts too small to measure?
   _____

Directions Look at the table of stars and their distances from Earth. List the stars in order of distance, from closest to farthest, on the lines.

10. _____
11. _____
12. _____
13. _____
14. _____
15. _____

Star	Distance from Earth
Aldebaran	646 trillion kilometers
Pollux	35 light-years
Vega	26 light-years
Alpha Centauri	41 trillion kilometers
Sirius	8.8 light-years
Capella	46 light-years

Earth Science

Workbook Activity 20
Chapter 5, Lesson 3

The Life of a Star

Directions The two flowcharts show the life cycles of two sizes of stars. Complete each flowchart by writing the correct star stage on each line. Use the definitions in parentheses to help you.

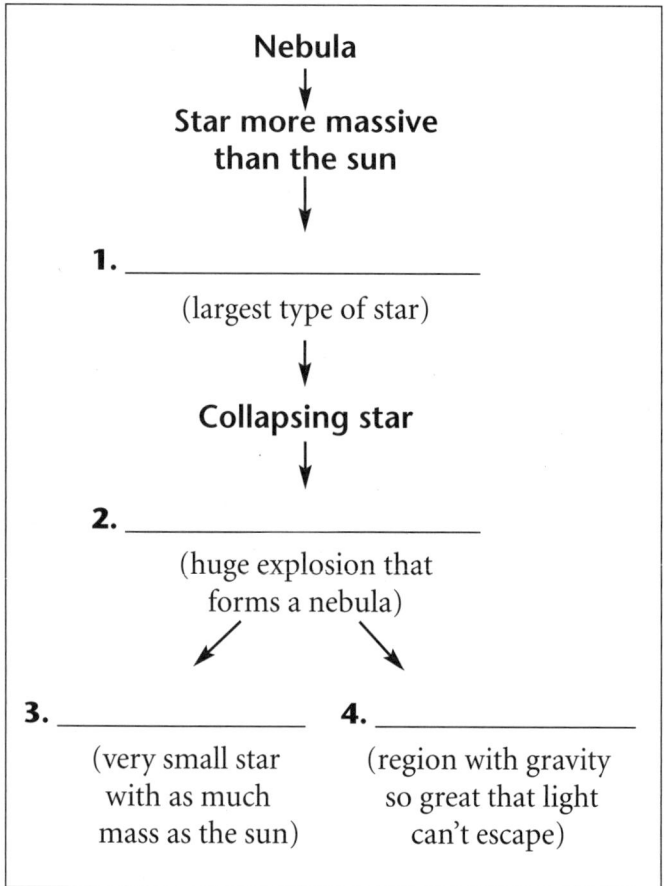

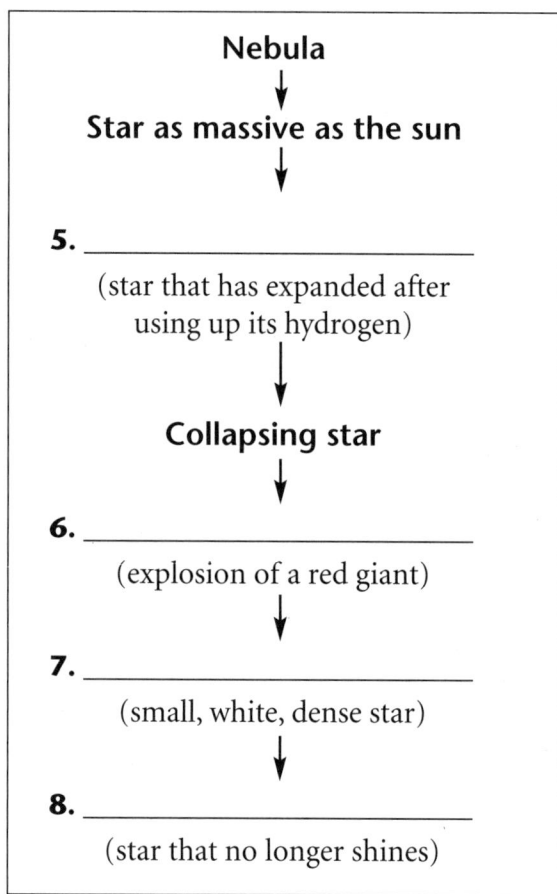

Directions Write the term from the Word Bank that best completes each sentence.

9. The life cycle of a star takes _____ of years.
10. A cloud of gas and dust drawn together by its own gravity is a(n) _____.
11. Gravity causes a star to _____.
12. Fusion causes a star to _____.
13. The outer layers of a star start to collapse when the star uses up its _____.
14. An example of a supergiant is _____.
15. A star that becomes a(n) _____ has a more dramatic end than a star that becomes a red giant.

Word Bank
Betelgeuse
billions
collapse
expand
hydrogen
nebula
supergiant

Earth Science

Name _____ Date _____ Period _____

Workbook Activity
Chapter 5, Lesson 4
21

Groups of Stars: Terms Review

Directions Read the clues to complete the puzzle.

Across

2. process that produces a star's energy
4. everything that exists
7. group of stars to which our solar system belongs (two words)
8. what may remain after a supernova (two words)
10. explosion of a supergiant

Down

1. patterns of stars seen from Earth
3. large group of stars
5. cloud of gas and dust in space
6. star that is larger than a red giant
9. explosion of a red giant

Directions Write the term from the Word Bank that best completes each sentence.

Word Bank				
black hole	elliptical	light-years	Polaris	the sun

11. The Milky Way galaxy is 100,000 _____ wide.
12. The star closest to Earth is _____.
13. The North Star is also called _____.
14. A(n) _____ galaxy is shaped like an oval.
15. The gravity of a(n) _____ is so great that light cannot escape.

Earth Science

Name _____ Date _____ Period _____ | **Workbook Activity**
Chapter 6, Lesson 1
22

Matter

Directions Write the term from the Word Bank that best completes each sentence.

Word Bank
chemical mass physical property state

1. All matter has _____ and takes up space.
2. Matter can be described by identifying its _____, or basic form.
3. A _____ is a characteristic that describes matter.
4. You can observe _____ properties of a substance without changing it.
5. To observe _____ properties, you must change the substance.

Directions Identify the state of each type of matter. Write *solid*, *liquid*, or *gas* on the line.

6. water _____
7. ice _____
8. steam _____
9. gasoline _____
10. air _____

Directions Name three physical properties and one chemical property of wood.

Matter: wood	
Physical Properties	**Chemical Property**
11. _____ 12. _____ 13. _____	14. _____

Directions Answer the question.

15. How are physical properties different from chemical properties?

_____

_____

Earth Science

Name _____ Date _____ Period _____ | **Workbook Activity** Chapter 6, Lesson 2 **23**

The Smallest Parts of Matter

Directions Match the term with its definition. Write the letter of the correct definition on the line.

_____ 1. proton

_____ 2. atom

_____ 3. nucleus

_____ 4. element

_____ 5. electron

A substance that cannot be changed or broken down into other substances

B center of an atom; contains protons and neutrons

C particle of an atom that moves outside the nucleus

D smallest particle of an element that keeps the element's properties

E particle found in the nucleus of an atom; makes each element unique

Directions Complete the table by writing the correct element name or symbol on the line. Use the numbered clues below the table to help you.

Common Elements in Earth's Rocks		
Element	Symbol	Number of Protons
6. _____	O	8
7. _____	Al	13
8. _____	Ca	20
sodium	9. _____	11
10. _____	H	1

6. This element combines with hydrogen to form water.
7. This shiny, lightweight metal is used to make ladders, kitchen utensils, and foil.
8. Dairy products, teeth, and bones contain this element.
9. This element combines with chlorine to make salt. The symbol comes from *natrium*, its Latin name.
10. This element combines with oxygen to form water.

Earth Science

Name _____ Date _____ Period _____

Workbook Activity
Chapter 6, Lesson 3
24

Compounds and Mixtures: Terms Review

Directions When you compare and contrast things, you tell how they are alike and different. Compare and contrast each pair of terms.

compound—element

1. Alike _____

_____

2. Different _____

_____

compound—mixture

3. Alike _____

_____

4. Different _____

_____

Directions Salt is sodium chloride, NaCl. A common type of sand is silicon dioxide, SiO_2. Use this information to answer the questions.

5. What elements make up salt? _____

6. In silicon dioxide, how many atoms of silicon combine with two oxygen atoms? _____

7. Are NaCl and SiO_2 examples of compounds or mixtures? _____

Directions Match each term with its definition. Write the letter of the correct definition on the line.

_____ **8.** atom **A** anything that has mass and takes up space

_____ **9.** chemical property **B** two or more substances mixed but not chemically combined

_____ **10.** compound **C** characteristic that describes matter

_____ **11.** matter **D** property observed when matter is chemically changed

_____ **12.** mixture **E** property observed without having to change matter

_____ **13.** nucleus **F** smallest unit of an element that keeps the element's properties

_____ **14.** physical property **G** part of an atom; contains neutrons and protons

_____ **15.** property **H** what forms when atoms of two or more elements combine chemically

Earth Science

Minerals

Directions Read the clues to complete the puzzle.

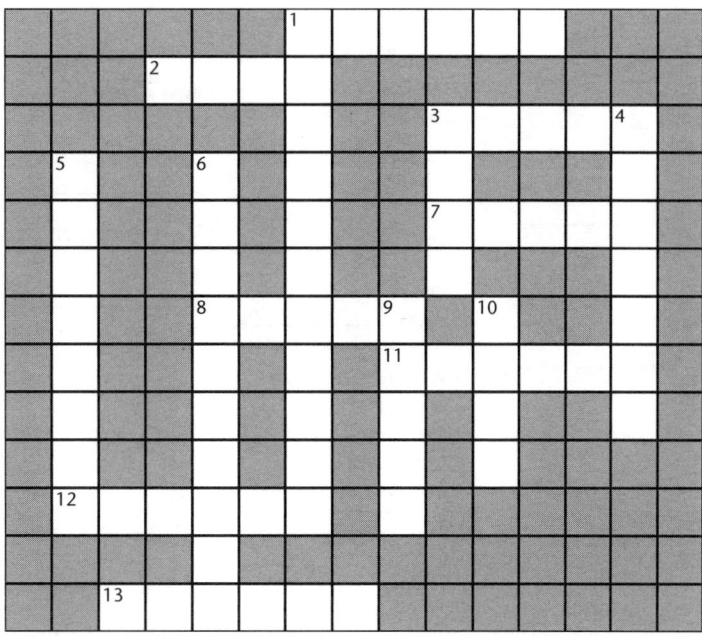

Across

1. Miners have to go down _____ to get to deep mines.
2. One way to separate minerals from rock is to _____ them in smelting ovens.
3. Minerals are often found in _____.
7. Geologists might test water from a _____ for traces of minerals.
8. Minerals are never _____.
11. Minerals have definite _____ patterns.
12. Minerals are not made of _____ things.
13. _____ is a common mineral made of silicon and oxygen.

Down

1. _____ involves skimming minerals in long patches off the earth's surface. (two words)
3. A few minerals are common, but most are _____.
4. Minerals near the _____ can be dug from pits.
5. A mineral has the same _____ makeup throughout.
6. _____ is the one continent where people don't mine minerals.
9. Minerals are formed naturally in the _____.
10. _____ is a precious mineral that is also an element.

Name	Date	Period	**Workbook Activity**
			Chapter 7, Lesson 2 **26**

Properties Used to Identify Minerals

Directions Look at the concept map of mineral properties. Match each empty box in the map with an answer from the Answer Bank. Write the letter of the correct answer in the box.

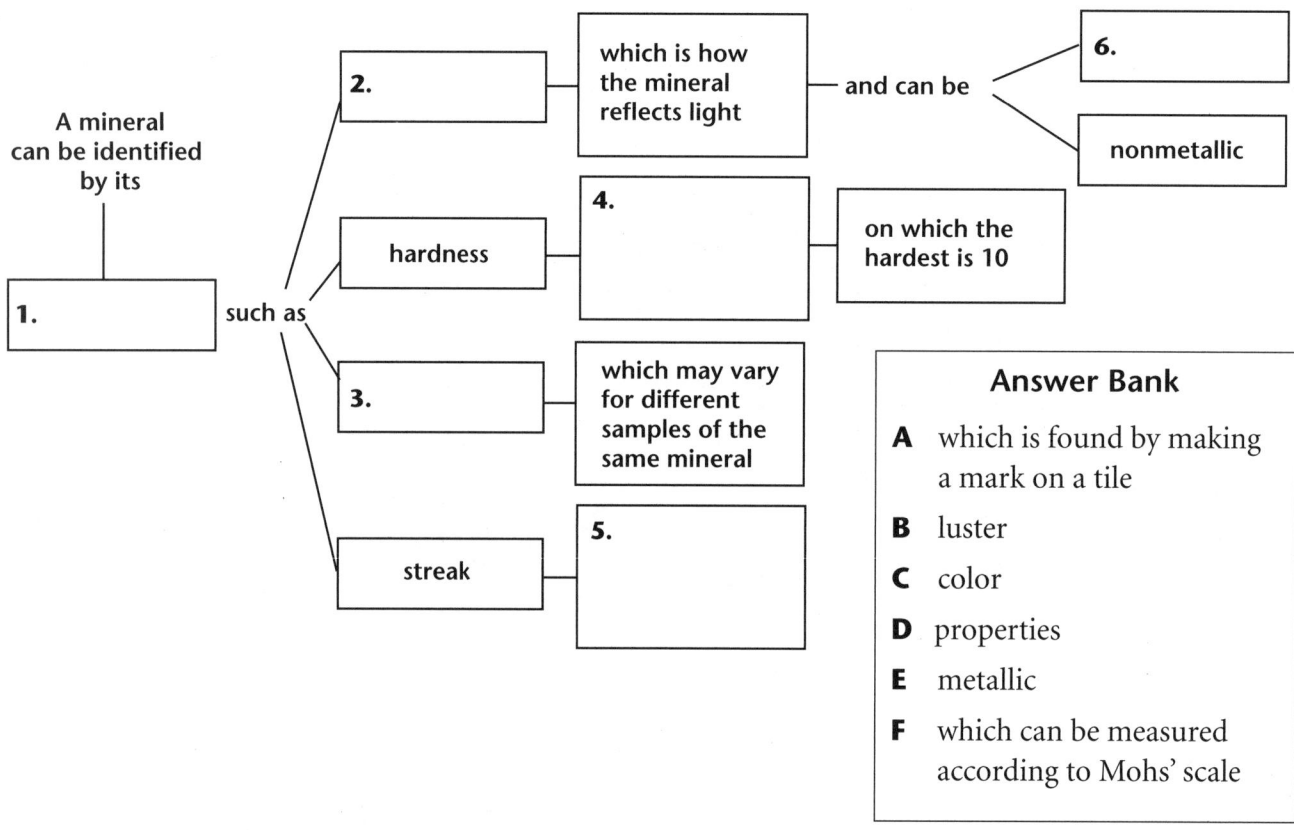

Directions Unscramble the word in parentheses to complete each sentence. Write the answer on the line.

7. _____ and pyrite may look alike, but their streaks are different. (dolg)

8. No other mineral is capable of scratching a _____. (donmaid)

9. One of the very softest minerals is _____. (lact)

10. Mohs' scale measures _____. (sendrash)

11. The _____ of quartz is described as glassy. (elrust)

12. A hard mineral can scratch a _____ one. (fots)

13. Sulfur is easy to recognize because it is bright _____. (wolley)

14. When doing a streak test, use a _____ tile. (thwie)

15. You can quickly test for hardness by using a copper _____. (nynep)

Earth Science

Name	Date	Period	**Workbook Activity**
			Chapter 7, Lesson 3 — 27

Other Physical Properties of Minerals

Directions Write the term from the Word Bank that best completes each sentence. Then circle each Word Bank term in the puzzle below.

1. _____ are the basic shapes that minerals tend to take.
2. The arrangements of a mineral's _____ give it a distinctive shape.
3. Some crystals are square, like little ice _____.
4. _____ is a common mineral many people use for cooking.
5. _____ crystals have six sides.
6. Very small minerals can only be seen with a(n) _____.
7. In outer _____, scientists can grow nearly perfect crystals.
8. When you strike a mineral, it will _____ in a specific way.
9. A mineral with _____ breaks along flat surfaces.
10. _____ is a jagged break pattern.
11. Specific gravity is a measure of a mineral's _____.
12. The _____ of a mineral compares its weight to the weight of water.
13. _____ has a specific gravity of 1.
14. A mineral with a specific gravity of 2 is _____ as heavy as water.
15. A person who designs and manufactures jewelry is a(n) _____.

Word Bank
atoms
break
cleavage
crystals
cubes
density
fracture
jeweler
microscope
quartz
salt
space
specific gravity
twice
water

```
M D K J G K Z U L R N Q P D S
I Z X Y E C L E A V A G E O P
C U B E S W E A R B O C O N E
R F I V B S E E R E N W O T C
O T O N O P G L N I T K L R I
S H Z T R A U Q E E R A I L F
C T L L R C A D A R S O W R I
O H E K Z E Q H D O M W D E C
P M L A L A X Y E R O R A K G
E B U E Y E P R N E T T Y O R
N A S R E C R Y S T A L S N A
T H E B I I T W I C B D A M V
O C K T L W I V T Z F A A V I
L S T I N T S E Y D W E N V T
I E N F R A C T U R E U C H Y
```

Earth Science

Name _____ Date _____ Period _____ | **Workbook Activity**
Chapter 7, Lesson 4 — **28**

Common Uses of Minerals: Terms Review

Directions Match each term with its definition. Write the letter of the correct definition on the line.

_____ 1. mineral   **A** color of the mark a mineral makes on a white tile

_____ 2. luster   **B** ability to split along flat surfaces

_____ 3. streak   **C** element or compound found in the earth

_____ 4. hardness   **D** ability of a mineral to resist scratches

_____ 5. crystal   **E** tendency to break with jagged edges

_____ 6. cleavage   **F** how a mineral reflects light

_____ 7. fracture   **G** basic shape that minerals tend to take

_____ 8. specific gravity   **H** mineral's weight compared to the weight of water

Directions Read the story below. Find a mineral from the Word Bank that is used to make each product. Write the mineral on the line.

Word Bank
bauxite   copper   corundum   gypsum   iron   quartz   talc

Last weekend Paolo built a partition in his basement. First he set up his

stepladder (**9.** _____). Then he fastened studs to

the floor and ceiling with three-inch nails (**10.** _____).

He replaced a window (**11.** _____). A friend helped him fit

the wallboard (**12.** _____) in place. He plastered any gaps

and holes and smoothed them with sandpaper (**13.** _____).

He finished the job with two coats of white paint (**14.** _____).

He furnished the room and plugged in the power cord (**15.** _____)

on his new TV. He was ready to relax.

Earth Science

Name _____ Date _____ Period _____ | **Workbook Activity** Chapter 8, Lesson 1 — **29**

Rocks and Rock Types

Directions 1–11. Write the terms from the Word Bank on the correct lines in the concept map below.

Word Bank					
basalt	geologists	igneous	metamorphic	pressure	sedimentary
cemented	gneiss	melted	minerals	sandstone	

1. studied by _____
2. made from _____

Rocks

three types

3. _____
4. _____
5. _____

6. layers are _____ together
7. changed by heat and _____
8. hot, _____ minerals harden

9. example: _____
10. example: _____
11. example: _____

Directions Answer each question.

12. How many minerals are found in the earth? _____
13. How many of these minerals make up most of the earth's rocks? _____
14. Which two types of rock can form on the earth's surface? _____
_____
15. Which two types of rock can form deep inside the earth? _____
_____

Earth Science

Name _____ Date _____ Period _____ | **Workbook Activity** Chapter 8, Lesson 2 | **30**

Igneous Rocks

Directions Write the word or phrase from the Word Bank that best completes each sentence.

1. Between 50 and 200 kilometers underground, temperatures are _____.
2. Hot, liquid rock is called _____.
3. The size of the mineral crystals in an igneous rock is its _____.
4. All igneous rocks form from _____.
5. Lava hardens to form _____ igneous rocks.

Word Bank
about 1,400°C
extrusive
hot, melted rock
magma
texture

Directions Write the terms from the Word Bank on the correct lines in the table below.

Word Bank
buildings   large   on the surface   small or none
granite   obsidian   sharp tools   under the ground

Features	Intrusive Rocks	Extrusive Rocks
where formed	6. _____	7. _____
size of crystals	8. _____	9. _____
example	10. _____	11. _____
use	12. _____	13. _____

Directions Answer each question.

14. Why do intrusive rocks have a coarse texture? _____

15. Why is obsidian glassy? _____

Earth Science

Sedimentary Rocks

Directions Read the clues to complete the puzzle.

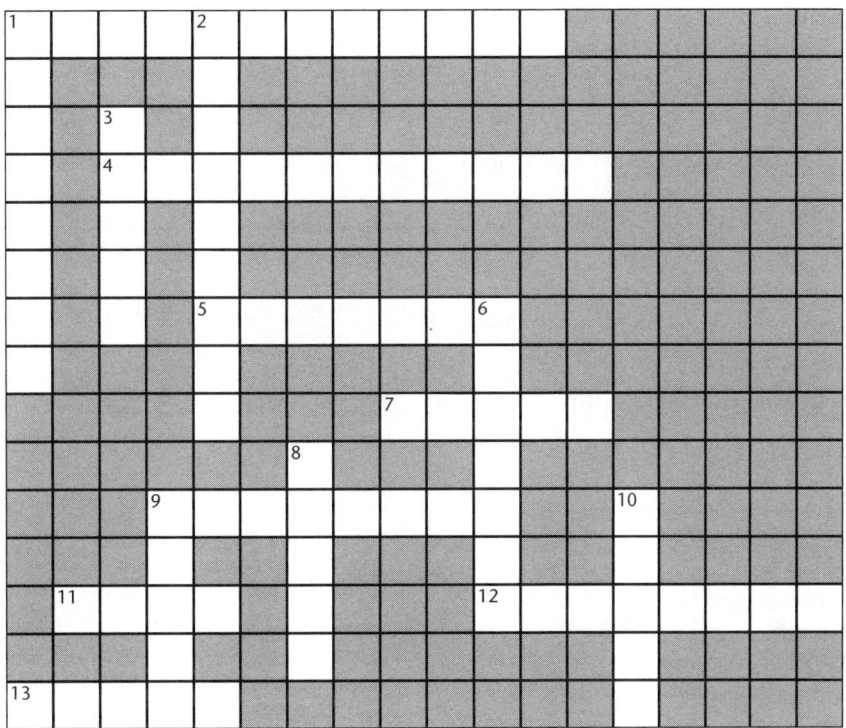

Across

1. clastic rock containing pebbles
4. Most ___ is produced by burning coal.
5. sedimentary rock made from living matter
7. type of organic limestone
9. solid particles that settle out of water
11. organic rock made from dead plants and animals
12. Layers of sediment are ___ together to make sedimentary rock.
13. Rock salt is left behind when salt ___ evaporates.

Down

1. sedimentary rock formed from dissolved chemicals
2. Layers of calcium carbonate become ___.
3. Sandstone might form from an ancient ___.
6. sedimentary rock made from bits of other rocks
8. Conglomerate might form where an ancient ___ was.
9. sedimentary rock made from tiny grains of mica and clay
10. Shale might form where an ancient ____ was.

Name _____ Date _____ Period _____ | Workbook Activity
Chapter 8, Lesson 4
32

Metamorphic Rocks

Directions Write the term from the Word Bank that best completes each sentence.

1. Metamorphic rocks form _____ the earth.

2–3. Metamorphic rock has been changed by _____ and _____.

4–5. Two types of metamorphic rock are _____ and _____.

6. Foliated rocks form in _____.

7. Slate can be split along its bands into _____.

8. Nonfoliated rocks usually form from _____ material.

9. In nonfoliated rocks, _____ combine and interlock to form harder rock.

Word Bank
bands
deep inside
foliated
heat
mineral crystals
nonfoliated
one
pressure
sheets of rock

Directions Write the terms from the Word Bank on the correct lines in the table below.

Word Bank					
columns	marble	no bands	slate	tiles	visible bands

Features	Foliated Rocks	Nonfoliated Rocks
description	10. _____	11. _____
example	12. _____	13. _____
use	14. _____	15. _____

Earth Science

The Rock Cycle: Terms Review

Directions Match each term with its definition. Write the letter of the correct definition on the line.

_____ 1. intrusive rock
_____ 2. metamorphic rock
_____ 3. rock
_____ 4. texture
_____ 5. igneous rock
_____ 6. chemical rock
_____ 7. sedimentary rock
_____ 8. lava
_____ 9. rock cycle
_____ 10. extrusive rock
_____ 11. sediment
_____ 12. foliated rock
_____ 13. clastic rock
_____ 14. organic rock
_____ 15. magma

A natural changes that cause one form of rock to become another
B size of crystals in igneous rock
C melted rock above the earth's surface
D igneous rock formed on the earth's surface (for example, obsidian)
E solid particles that settle out of liquid
F sedimentary rock formed from bits of other rocks
G rock formed by intense heat and pressure
H metamorphic rock with bands
I rock formed as melted rock cools and hardens
J hot, liquid rock inside the earth
K sedimentary rock formed from chemicals dissolved in water
L natural, solid mixture of minerals
M rock formed by cementing layers of sediment
N igneous rock formed underground from cooled magma
O rock formed from the remains of living things

Directions The Word Bank lists some forces that change rocks. Write the terms from the Word Bank on the correct lines in the diagram.

Word Bank
compacting and cementing
cooling and hardening
heat and pressure
melting
weathering and erosion

20. _____
19. _____
16. _____
17. _____
18. _____

Earth Science

Name _____ Date _____ Period _____ | **Workbook Activity** **34**
Chapter 9, Lesson 1

Gases in the Atmosphere

Directions 1–10. Put the steps of each cycle in the correct order. Write the numbers 1 to 6 on the lines. (1 is already marked for you.)

The Oxygen-Carbon Dioxide Cycle

_____ They release carbon dioxide into the atmosphere.

__1__ Animals and people breathe in air.

_____ Plants release oxygen into the air.

_____ Plants take in the carbon dioxide through their leaves.

_____ Their bodies use oxygen to change food into energy.

_____ Plants use carbon dioxide to make sugar.

The Nitrogen Cycle

_____ Animals and people take in nitrogen when they eat plants or plant-eating animals.

_____ Bacteria in the soil break down these wastes.

__1__ Bacteria in the soil change nitrogen gas into chemical compounds.

_____ Nitrogen is returned to the soil as animal waste and as dead plants and animals.

_____ Bacteria release nitrogen into the air and into the soil.

_____ Plants take in these nitrogen-containing compounds through their roots.

Directions Answer the questions.

11. What is the atmosphere? _____

_____

12. What percent of the atmosphere is oxygen gas? _____

13. What percent of the atmosphere is nitrogen gas? _____

14. Compare how plants take in nitrogen with how they take in carbon dioxide.

_____

_____

15. Compare how animals take in nitrogen with how they take in oxygen.

_____

_____

Earth Science

Name _____ Date _____ Period _____

Workbook Activity
Chapter 9, Lesson 2
35

Layers of the Atmosphere

Directions 1–8. Look at the diagram of the atmosphere. Then write the names of the numbered layers in the first column of the table below. Complete the table by writing a short description of each layer.

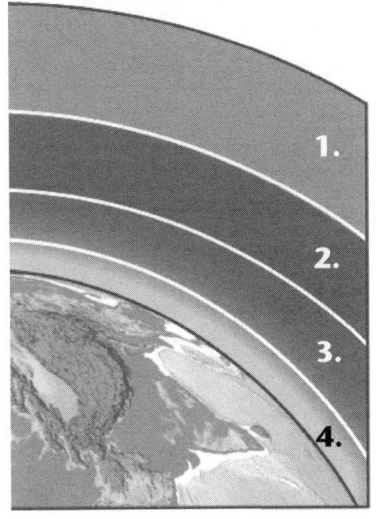

Name of Layer	Description
1.	5.
2.	6.
3.	7.
4.	8.

Directions Answer the questions.

9. Why does the troposphere contain 75 percent of the air particles in the atmosphere?

_____

_____

10. Why do mountain climbers often need oxygen tanks when they climb?

_____

_____

11. Where is the ozone layer? Why is it important? _____

_____

12. In which layer of the atmosphere is the air thinnest? _____

13. In which layer of the atmosphere is most weather activity? _____

14. Where are ions found in the atmosphere? How are ions formed? _____

_____

15. Why are radio messages transmitted better at night? _____

_____

Earth Science

Name _____ Date _____ Period _____ | **Workbook Activity** Chapter 9, Lesson 3 | **36**

Clouds

Directions Write the term from the Word Bank that best completes each sentence. Then find and circle each term in the puzzle.

1. A(n) _____ is a mass of tiny water droplets.
2. The height above the earth's surface is _____.
3. Some clouds form when air is forced up a(n) _____.
4. _____ is a cloud that forms near the ground.
5. There are _____ main types of clouds.
6. Clouds are grouped according to their _____ and altitude.
7. Fluffy _____ clouds are found between 2,000 and 7,000 meters.
8. High, thin, wispy clouds are called _____ clouds.
9. Heat from the sun causes water to _____, or change into gas.
10. When air cools, water vapor in it may _____, or change into liquid.
11. Air contains a gas called water _____.
12. Cirrus and cumulus clouds are often seen in _____ weather.
13. Low, flat clouds that often bring rain are _____ clouds.
14. Fog usually forms in early morning in low areas or over warm _____.
15. Cirrus clouds are made of _____ instead of water droplets.

Word Bank
altitude
cirrus
cloud
condense
cumulus
evaporate
fair
fog
ice crystals
mountain
shape
stratus
three
vapor
water

```
R R O A V F B A N T V A P O R
P B B F L P Z D G N F Z E Z L
C L V E O L D R R M S S E T J
S U A L T I T U D E U T K H E
E T M T H R E E O R U D E R C
S O R U F A I R R Z F M E O L
R N X A L D A I T K O N V T O
C H I L T U C T T N I I A A U
O W O V A U S R P A Y H P A D
N A F F W S S I T P P I O D W
D T M O L J T N W T E A R E E
E E N G E J U B R F P Y A F L
N R N T P O F W R R A U T C U
S Y W X M E C E G T H W E A T
E U G O I C E C R Y S T A L S
```

Earth Science

Name	Date	Period	**Workbook Activity**
			Chapter 9, Lesson 4 — 37

Precipitation

Directions 1–8. Complete the table using answers from both Answer Banks. Write the letter of the correct answer in each numbered space in the table.

Kind of Precipitation	What It Is	How It Forms
rain	1.	5.
hail	2.	6.
sleet	3.	7.
snow	4.	8.

Answer Bank for "What It Is" Column

- **A** freezing rain
- **B** ice crystals
- **C** droplets of water
- **D** balls of ice

Answer Bank for "How It Forms" Column

- **E** Raindrops fall through a layer of cold air and freeze into ice particles. The ice particles may form as the raindrops hit a surface or before they hit the ground.
- **F** Water droplets or ice crystals collide in clouds and combine. Then they fall through above-freezing air.
- **G** Ice crystals in clouds in middle and high latitudes combine until they are heavy enough to fall. They fall through below-freezing air.
- **H** Strong winds toss ice crystals up and down in tall cumulus clouds. The clouds are below freezing at the top, but above freezing near the bottom. Layers of water freeze around the ice crystals before they fall.

Directions Write a short answer for each question.

9. What is precipitation? _____

10. Where above the earth are most stratus clouds made of water droplets? _____

11. Where above the earth are most clouds made of ice crystals? _____

12. Describe what happens in an ice storm. _____

13. How are sleet and snow alike? _____

14. How are sleet and snow different? _____

15. What determines whether precipitation falls as a liquid or a solid? _____

Earth Science

Wind Patterns: Terms Review

Directions Read the clues to complete the puzzle.

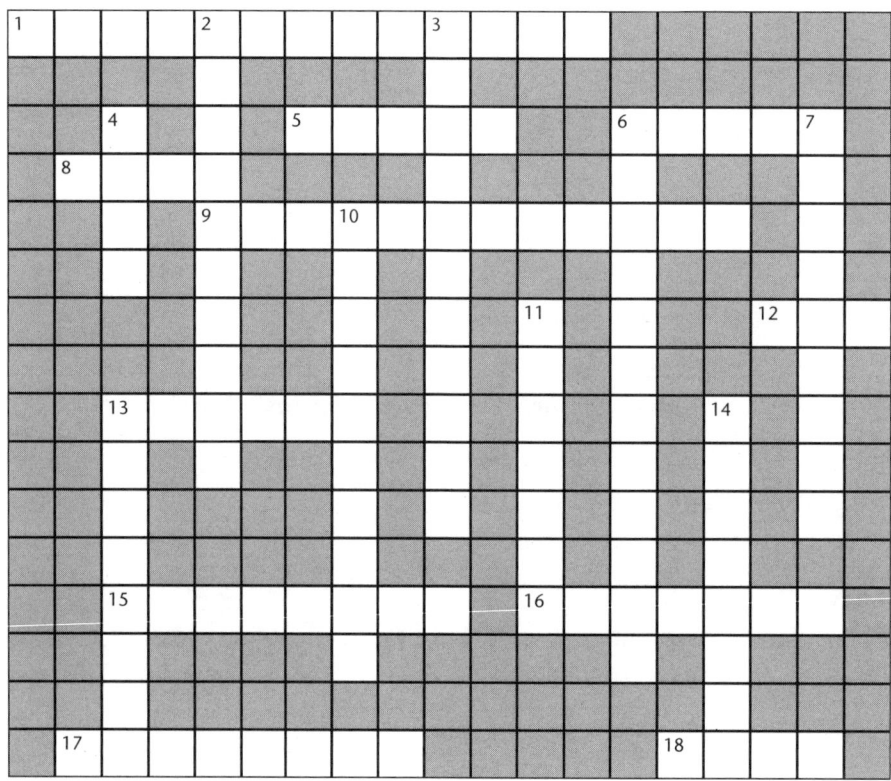

Across

1. moisture falling to the earth
5. water ___: water in a gas phase
6. ___ wind: strong wind just north or south of the equator
8. ___sphere: coldest layer of the atmosphere
9. atmosphere layer that includes ozone
12. cloud near the ground
13. thin, wispy type of cloud
15. polar ___: wind near a pole that brings cold, stormy weather
16. low, flat cloud type
17. prevailing ___: wind blowing from the west
18. wind ___: pattern of wind movement around the earth

Down

2. section of the atmosphere with electrically charged particles
3. bottom layer of the atmosphere
4. wind ___: cycle of air flow
6. outermost layer of the atmosphere
7. to change from a liquid to a gas
10. layer of gases around the earth
11. puffy, white cloud type
13. to change from a gas to a liquid
14. height above the earth's surface

Name _____ Date _____ Period _____ | Workbook Activity
Chapter 10, Lesson 1
39

Weather Conditions and Measurements

Directions In the table, write the name of the instrument next to the weather condition it measures.

Name of Instrument	What It Measures
1.	air temperature
2.	air pressure
3.	relative humidity
4.	wind speed
5.	wind direction
6.	amount of rain

Directions Answer the questions.

7. What is weather? _____
_____

8. How are a Fahrenheit thermometer and a Celsius thermometer different?
_____
_____

9. How does an aneroid barometer work? _____
_____

10. What is relative humidity? _____
_____
_____

11. If the front end of a wind vane is pointing west and the back end is pointing east, which way is the wind blowing? _____

12. What are three ways to measure precipitation? _____
_____

13. A wind is blowing from west to east. Is it a west wind or an east wind? _____

14. The barometric pressure rises from 74 centimeters to 76 centimeters. What kind of weather may be coming? _____

15. Water freezes and ice melts at 32°F. What is this temperature in Celsius? _____

Earth Science

Name	Date	Period	**Workbook Activity**
			Chapter 10, Lesson 2
			40

Weather Patterns and Predictions

Directions Match each term with its description. Write the letter of the correct description on the line.

_____ 1. cold front   **A** section of air having same temperature and humidity

_____ 2. low   **B** area of high pressure

_____ 3. warm front   **C** line connecting areas of equal air pressure

_____ 4. front   **D** cold air mass pushing out and under warm air mass

_____ 5. isobar   **E** warm air mass gliding up and over cooler air mass

_____ 6. air mass   **F** area of low pressure

_____ 7. high   **G** boundary line between two air masses

Directions Label each diagram with one of these terms: *low*, *high*, *cold front*, or *warm front*. On the lines below each diagram, describe the weather conditions that are represented. Consider temperature, precipitation, wind, sky conditions, and air pressure.

8. _____

9. _____
_____
_____
_____

10. _____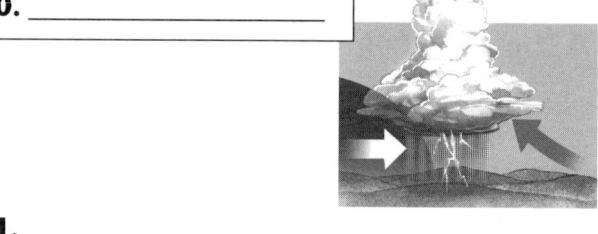

11. _____
_____
_____
_____

12. _____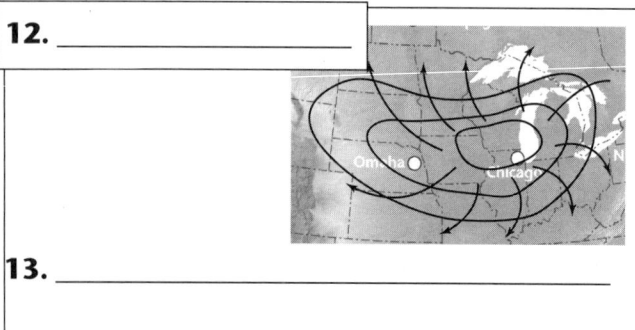

13. _____
_____
_____
_____

14. _____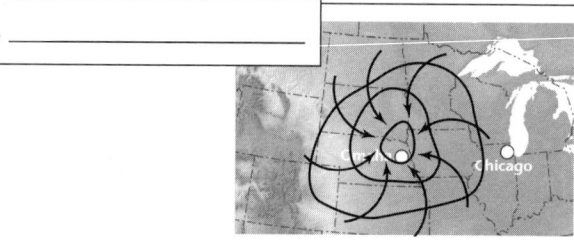

15. _____
_____
_____
_____

Earth Science

Name _____ Date _____ Period _____

Workbook Activity
Chapter 10, Lesson 3
41

Storms

Directions Write the word from the Word Bank that best completes each sentence. Then find and circle each word in the puzzle below.

1. A whirling, funnel-shaped cloud formed in a thunderstorm is a(n) _____.

2. Large, dark clouds that produce heavy rain are _____.

3. A(n) _____ occurs when warm air is forced up by a cold front.

4. _____ forms in clouds when electrical current passes between negative and positive charges.

5. When lightning occurs, _____ is the sound made by the suddenly expanding, heated air.

6. Tornadoes are likely to occur over flat areas such as _____.

7. A(n) _____ is the largest kind of storm.

8. You see lightning _____ you hear thunder.

9. Air pressure in a tornado is very _____.

10. When a hurricane moves inland, it drops a lot of _____.

11. A force that slows hurricanes on land is _____.

12. As it moves over land, a hurricane loses force because it has no source of heat and _____.

13. The calm area at the center of a hurricane is called the _____.

14. _____ are caused by rapid changes in the movement of air masses.

15. Hurricanes form in the tropics near the _____.

Word Bank
before
equator
eye
friction
hurricane
lightning
low
moisture
prairies
rain
storms
thunder
thunderheads
thunderstorm
tornado

A	B	S	E	I	R	I	A	R	P	Q	W	T	I
C	E	V	Z	Y	D	B	P	U	T	Y	E	H	N
T	N	B	S	T	E	E	C	K	B	O	G	U	D
H	A	F	R	B	L	F	U	D	P	C	N	N	T
U	K	R	A	I	N	O	W	A	E	J	I	D	A
N	Y	I	O	T	C	R	Y	T	S	H	N	E	R
D	E	C	A	U	K	E	Q	L	V	U	T	R	P
E	U	T	R	T	H	U	N	D	E	R	H	S	N
R	V	I	Y	O	V	W	R	J	U	R	G	T	W
H	K	O	S	R	P	O	S	A	O	I	I	O	B
E	S	N	L	N	T	Q	M	B	W	C	L	R	E
A	T	B	O	A	E	C	R	Y	D	A	P	M	O
D	C	D	U	D	N	J	O	T	J	N	C	V	C
S	W	Q	M	O	I	S	T	U	R	E	K	A	W
N	E	O	V	D	R	E	S	K	R	A	U	Y	S

Earth Science

Name _____ Date _____ Period _____ | **Workbook Activity**
Chapter 10, Lesson 4 — 42

World Climates: Terms Review

Directions Write the answer to the question in each box.

What are the three major climate zones from coldest (1) to warmest (3)?	What are three factors that affect climate?
1. _____	4. _____
2. _____	5. _____
3. _____	6. _____

Directions Match each term with its definition. Write the letter of the correct definition on the line.

_____ 7. relative humidity

_____ 8. hurricane

_____ 9. tornado

_____ 10. front

_____ 11. isobar

_____ 12. high

_____ 13. climate

_____ 14. air pressure

_____ 15. weather

_____ 16. low

_____ 17. air mass

A state of the atmosphere at a given time and place

B amount of water vapor in air compared to the maximum amount of water vapor air can hold

C section of air with the same temperature and humidity

D line on a weather map connecting areas of equal pressure

E tropical storm with high winds revolving around an eye

F powerful wind storm with whirling, funnel-shaped cloud and very low pressure

G cold area of high air pressure

H warm area of low air pressure

I moving boundary line between two air masses

J force of air against a unit of area

K average weather pattern of a region over a long time

Directions Write the weather condition that each instrument measures.

18. anemometer _____

19. psychrometer _____

20. barometer _____

Earth Science

Name	Date	Period	**Workbook Activity**
			Chapter 11, Lesson 1 43

The Water Cycle

Directions Look at the diagram of the water cycle on page 256 in your textbook. Then explain the water cycle by writing a description of each term below.

1. evaporation _____

_____

2. condensation _____

_____

3. precipitation _____

_____

4. runoff _____

_____

5. groundwater _____

_____

Directions Answer the questions.

6. How does the sun power the water cycle? _____

_____

_____

7. If precipitation doesn't sink into the ground, what two things can happen to it?

_____

_____

8. Suppose you lived in a hilly area with a lot of runoff. What could you do to help the ground soak up more water?

_____

_____

9. Where does most runoff eventually end up? _____

10. Could a farmer near the sea coast use salt water to water crops? Why or why not?

_____

_____

Earth Science

Name _____ Date _____ Period _____ | **Workbook Activity**
Chapter 11, Lesson 2 — **44**

Sources of Fresh Water

Directions Match each term with its description. Write the letter of the correct description on the line.

_____ 1. divide   A forms when the roof of a cave collapses

_____ 2. drainage basin B top of the groundwater layer

_____ 3. geyser   C river that joins another river

_____ 4. porous   D made by constructing a dam across a river

_____ 5. reservoir   E hot groundwater and steam shooting into the air

_____ 6. sinkhole   F land area drained by a river and its tributaries

_____ 7. spring   G having many spaces for water and air to flow

_____ 8. tributary   H separates two drainage basins

_____ 9. water table   I groundwater flowing naturally out of the ground

Directions Write an answer to each question.

10. What is the water table? _____

_____

11. How does a sinkhole form? _____

_____

12. Name three bodies of fresh water that are above the ground.

_____

_____

_____

13. What percent of the earth's water is fresh water? _____

14. How do lakes gain and lose water? _____

_____

_____

15. What are three purposes of reservoirs? _____

_____

_____

_____

Earth Science

Name _____ Date _____ Period _____ | Workbook Activity Chapter 11, Lesson 3 | 45

Oceans: Terms Review

Directions Write the term from the Word Bank that best completes each sentence.

Word Bank				
benthos	divide	mid-ocean ridge	runoff	thermocline
continental shelf	drainage basin	nekton	salinity	trench
continental slope	geysers	porous	seamount	tributary
currents	groundwater	reservoir	sinkhole	water cycle
				water table

1. Precipitation that sinks into the ground becomes _____.
2. Precipitation that does not sink into the ground or evaporate becomes _____.
3. Water moving between the atmosphere and the earth's surface is the _____.
4. A river that joins another river is a _____.
5. When the roof of a cave collapses, a _____ forms.
6. The top of the groundwater layer is the _____.
7. Water can sink into the ground because the soil is _____.
8. The land area in which runoff drains into a large river is a _____.
9. An artificial lake made by building a dam is a _____.
10. A ridge that separates drainage basins is a _____.
11. The saltiness of water is _____.
12. A deep valley in the ocean floor is a _____.
13. A mountain chain on the ocean floor is a _____.
14. The temperature drops sharply in the _____.
15. The part of a continent that extends underwater is the _____.
16. A _____ dips from a continental shelf down to the ocean floor.
17. An underwater mountain that is often a volcano is a _____.
18. Ocean life includes plankton, _____, and _____.
19. Winds cause up-and-down waves as well as flowing ocean streams called _____.
20. Groundwater flows out of springs and blasts out of _____.

Earth Science

Name _____ Date _____ Period _____ | **Workbook Activity** Chapter 12, Lesson 1 | **46**

Weathering

Directions Look at the diagram of soil layers. Then name and describe each numbered layer in the table.

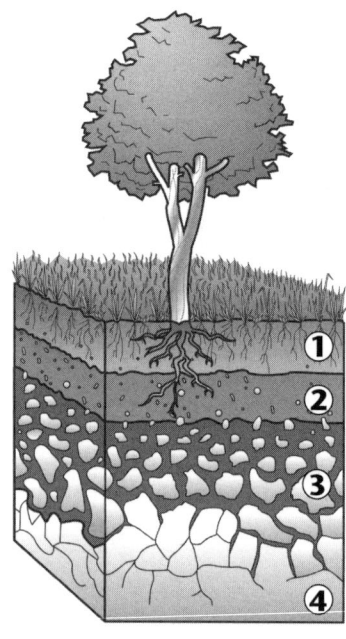

Layer Name	Description
1.	5.
2.	6.
3.	7.
4.	8.

Directions Match the term with its definition. Write the letter of the correct definition on the line.

_____ 9. weathering

_____ 10. mechanical weathering

_____ 11. chemical weathering

_____ 12. soil

A breaking down of rocks without changing their chemical makeup

B mixture of rock pieces and the remains of dead plants and animals

C breaking down of rocks on the earth's surface

D breaking down of rocks by changing their chemical makeup

Directions Write the answer to each item below.

13. Name three factors in the environment that cause weathering. _____

_____

14. Describe two examples of mechanical weathering. _____

_____

_____

15. Describe two examples of chemical weathering. _____

_____

Earth Science

Name _____ Date _____ Period _____ | **Workbook Activity**
Chapter 12, Lesson 2
47

Erosion Caused by Water

Directions Compare and contrast each pair of terms.

weathering—erosion

1. Alike _____

2. Different _____

valley—canyon

3. Alike _____

4. Different _____

meander—oxbow lake

5. Alike _____

6. Different _____

delta—alluvial fan

7. Alike _____

8. Different _____

wave erosion—wave deposition

9. Alike _____

10. Different _____

young river—mature river

11. Alike _____

12. Different _____

Directions Answer the questions.

13. How does river water erode the land? _____
_____

14. What part does sediment play in river erosion and deposition? _____
_____

15. How does a river valley change with time? _____
_____

Earth Science

Name _____ Date _____ Period _____ | **Workbook Activity** Chapter 12, Lesson 3 | **48**

Erosion Caused by Glaciers

Directions Match each term with its definition. Write the letter of the correct definition on the line.

_____ 1. continental glacier   **A** jagged peak formed by intersecting cirques

_____ 2. glacier   **B** body of water left by a glacier

_____ 3. alpine glacier   **C** glacier that begins in the mountains

_____ 4. cirque   **D** ridge of sediment left by a glacier

_____ 5. horn   **E** thick mass of ice covering a large area

_____ 6. moraine   **F** piece of a glacier that breaks off in water

_____ 7. iceberg   **G** bowl-shaped basin carved by a glacier

_____ 8. glacial lake   **H** glacier covering a broad area of land near a pole

Directions Write a short answer to each question.

9. What causes glaciers to move? _____

10. How do glaciers erode the land underneath them? _____

11. How does an alpine glacier change the shape of a valley? _____

12. How did the Great Lakes form? _____

13. How do we know if a lake was made by a glacier? _____

14. About 10 percent of the earth's surface is covered with glaciers today. Where are most of these glaciers found? _____

15. What two conditions are needed for a glacier to form? _____

Earth Science

Erosion Caused by Wind and Gravity: Terms Review

Directions
Read the clues to complete the puzzle.

Across

1. dropping of eroded sediment
3. ridge left by a glacier
5. process of breaking rocks apart
9. basin carved out by an alpine glacier
10. glacier in a mountainous region
12. weathering in which chemical makeup stays the same
15. where a river flows into a larger body of water
16. mixture of rock bits and the remains of dead plants or animals
17. layer of soil containing minerals from topsoil
18. area that a river covers when it overflows its banks

Down

1. fan-shaped land area where a river enters a lake or ocean
2. C-shaped lake formed from a meander
4. process in which rust forms
6. moving of weathered rock and soil
7. thick mass of ice covering a large area
8. weathering that changes the makeup of a rock
11. fan-shaped land deposited at the base of a mountain
12. looping curve in a river
13. peak formed where cirques come together
14. richest soil layer

Name _____ Date _____ Period _____

Workbook Activity
Chapter 13, Lesson 1
50

Movement of the Earth's Crust

Directions Write each phrase from the Answer Bank under the part of the earth it describes.

Answer Bank		
8 to 70 kilometers thick	innermost layer	iron and nickel
2,900 kilometers thick	outer layer	churning, hot rock
3,500 kilometers thick	middle layer	continents and ocean floor

Crust

1. _____
2. _____
3. _____

Mantle

4. _____
5. _____
6. _____

Core

7. _____
8. _____
9. _____

Directions Write the terms from the Word Bank that best complete the sentences.

10. Alfred Wegener believed all of the continents on the earth were joined millions of years ago as one landmass called _____.

11. The surface of the earth's crust is made of large _____, which _____, _____, or slide past each other.

12. The earth's plates move because the magma beneath them has _____ currents, or circular motion.

Word Bank
collide
convection
move apart
Pangaea
plates

Directions Explain each theory on the lines below.

13. **continental drift** _____
_____
_____

14. **sea-floor spreading** _____
_____
_____

15. **plate tectonics** _____
_____
_____

Earth Science

Name _____ Date _____ Period _____

Workbook Activity
Chapter 13, Lesson 2
51

Volcanoes

Directions Write each answer from the Answer Bank in the correct part of the table.

Answer Bank

both explosions and quiet flows	loose rock particles	small and steep
explosive blasts	low and broad	tall
lava layers and rocky layers	quiet flows	thin lava layers

Type of Volcano	Type of Eruption	Type of Material	Shape
cinder cone	1.	2.	3.
shield	4.	5.	6.
composite	7.	8.	9.

Directions Write the term from the Word Bank that best completes each sentence.

10. Volcanoes form because magma rises through openings called _____.

11. Most volcanoes form where two _____ meet.

12. A small volcano built of loose rock particles is called a(n) _____.

13. Shield volcanoes form from thin basalt lava, so their eruptions are not very _____.

14. Composite volcanoes form from alternating explosive eruptions and _____ eruptions of lava.

15. _____ volcanoes have a broad shape.

Word Bank
cinder cone
explosive
plates
quiet
shield
vents

Earth Science

Name _____ Date _____ Period _____

Workbook Activity
Chapter 13, Lesson 3
52

Mountains

Directions Write each answer from the Answer Bank in the correct part of the table.

Answer Bank
• Cascade Range • folding
• Continental plates collide, bending rock layers. • Grand Tetons
• Himalayas
• The earth's crust breaks, and blocks of rock rise. • One plate sinks beneath another or two plates separate.
• fault • volcanic

Ways Mountains Form		
Type of Formation	Force Causing Formation	Example
1.	2.	3.
4.	5.	6.
7.	8.	9.

Directions 10–15. Draw lines to connect each fault diagram with its name on the left and its description on the right.

strike-slip fault   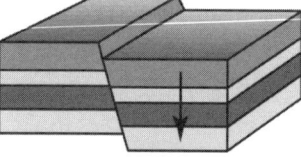   Overhanging block of rock is raised.

normal fault   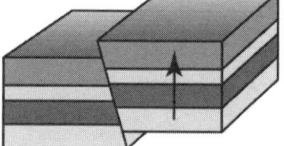   Blocks of rock slide past each other.

reverse fault   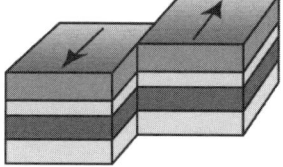   Overhanging block of rock slides down.

Earth Science

Name _____ Date _____ Period _____

Workbook Activity
Chapter 13, Lesson 4
53

Earthquakes: Terms Review

Directions Draw lines to connect each earthquake wave with its description.

1. P-wave   fastest, causes rocks to vibrate back and forth
2. L-wave   slower, causes rocks to vibrate up and down
3. S-wave   slowest, causes the ground to twist and bend

Directions Match each term with its definition. Write the letter of the correct definition on the line.

___ 4. focus   **A** theory that the earth's landmasses move
___ 5. epicenter   **B** circular motion of a gas or liquid as it heats
___ 6. Pangaea   **C** earthquake's origin inside the earth
___ 7. seismograph   **D** mountain formed when magma erupts
___ 8. volcano   **E** shaking of the earth's crust
___ 9. plate tectonics   **F** theory that the earth's crust is made of moving sections
___ 10. tsunami   **G** large ocean wave caused by an earthquake
___ 11. folding   **H** point directly over the focus of an earthquake
___ 12. sea-floor spreading **I** theory that new crust forms at mid-ocean ridges
___ 13. continental drift   **J** name of the scale that measures earthquake strength
___ 14. vent   **K** instrument that detects earthquake waves
___ 15. convection current   **L** opening at the top of a volcano
___ 16. earthquake   **M** process in which rock layers bend under pressure
___ 17. Richter   **N** single landmass that separated into continents

Directions Write the terms from the Word Bank in the correct boxes below.

Word Bank				
cinder cone	core	mantle	reverse	strike-slip
composite	crust	normal	shield	

18. Faults	**19. Volcanoes**	**20. The Earth's Layers**
• _____	• _____	• _____
• _____	• _____	• _____
• _____	• _____	• _____

Earth Science

Workbook Activity
Chapter 14, Lesson 1
54

The Rock Record

Directions Find the lettered phrase that best completes each sentence. Write the letter of the correct phrase on the line.

_____ 1. To find out about the earth's past, scientists ___.
_____ 2. Because of fossils, we know that ___.
_____ 3. Organisms that are buried quickly after death ___.
_____ 4. When animals become trapped in tree sap, their ___.
_____ 5. Hard ___ do not decay easily and may become fossils.
_____ 6. All of the time since the earth's formation is ___.
_____ 7. Petrification occurs when minerals replace ___.
_____ 8. A fossil is the remains of an organism ___.
_____ 9. A mold is the ___ where an organism was buried.
_____ 10. If minerals fill a mold, ___.

A certain organisms once existed
B space left in a rock
C study rock layers and fossils
D can become fossils
E a cast forms
F geologic time
G a buried organism
H preserved in the earth's crust
I teeth, bones, and shells
J actual bodies can be preserved

Directions Match each fossil description with one of the three terms below. Write P, I, or T on the line.

| P petrification | I imprint (mold or cast) | T trapped and preserved |

_____ 11. A mosquito is caught in amber.
_____ 12. A trilobite decays, leaving an imprint.
_____ 13. A saber-toothed tiger falls in a tar pit.
_____ 14. The wood in a tree branch is replaced by minerals.
_____ 15. Minerals fill the mold of a seashell.
_____ 16. A log turns into minerals.
_____ 17. A leaf decays and leaves its shape in the sediment.
_____ 18. A seed is covered with sap.
_____ 19. A wooly mammoth freezes in a snowstorm.
_____ 20. An imprint of fish bones is seen in a rock.

Earth Science

Name _____ Date _____ Period _____

Workbook Activity
Chapter 14, Lesson 2
55

The Ages of Rocks and Fossils

Directions Compare and contrast the two terms below. Explain how they are alike and how they are different.

relative dating—absolute dating

1. How are they alike? _____

_____

_____

2. How are they different? _____

_____

_____

Directions Answer the questions.

3. Give an example of an index fossil and tell how scientists might use such a find.

_____

_____

4. What is the principle of superposition? _____

_____

5. What is the principle of crosscutting relationships? _____

_____

6. How does the half-life of a radioactive element help determine the age of a rock or fossil?

_____

_____

Directions Identify each term below with a method of dating rocks. On the line, write *R* for relative dating or *A* for absolute dating.

7. carbon-14 _____
8. radioactive element _____
9. half-life _____
10. determining actual age _____
11. comparing layers _____
12. crosscutting _____
13. uranium-238 _____
14. index fossil _____
15. superposition _____

Earth Science

Name _____ Date _____ Period _____

Workbook Activity
Chapter 14, Lesson 3
56

Eras in the Geologic Time Scale: Terms Review

Directions Match each term with its description. Write the letter of the correct description on the line.

_____ 1. principle of superposition
_____ 2. Paleozoic Era
_____ 3. geologic time scale
_____ 4. fossil
_____ 5. radioactive element
_____ 6. index fossil
_____ 7. principle of crosscutting relationships
_____ 8. petrification
_____ 9. half-life
_____ 10. Mesozoic Era

A era marked by trilobites and other sea life
B youngest feature cuts across other rock layers
C fossil that provides clues to the age of a rock
D element that decays to form another element
E outline of the earth's history
F dinosaur era
G oldest rock layer is on the bottom
H minerals replace a buried organism
I preserved remains of an organism
J length of time for half of an element's atoms to decay

Directions 11–18. In each box below, write the numbers 1 to 4 on the lines to show the correct order in geologic history.

_____ Paleozoic Era
_____ Cenozoic Era
_____ Precambrian Era
_____ Mesozoic Era

_____ Swamps form; coal begins to develop.
_____ The earth's crust and mantle form.
_____ The Rocky Mountains form.
_____ The Great Lakes form after the last ice age.

Directions Contrast each pair of terms. Explain how the terms in each pair are different.

19. absolute dating—relative dating _____

_____

_____

20. cast—mold _____

_____

_____

Earth Science